Math Mammoth Grade 4 Tests and Cumulative Revisions

for the complete curriculum
(International Series)

Includes consumable student copies of:

- Chapter Tests
- End-of-year Test
- Cumulative Revisions
- Fraction Cut-outs

By Maria Miller

Contents

Grade 4, Chapter 1

End-of-Chapter Test

Instructions to the student:

Do not use a calculator. Answer each question in the space provided.

Instructions to the teacher:

My suggestion for grading the chapter 1 test is below. The total is 15 points. Divide the student's score by the total of 15 to get a decimal number, and change that decimal to percent to get the student's percentage score.

Question #	Max. points	Student score
1	1 points	
2	3 points	
3	3 points	
4	2 points	

Question #	Max. points	Student score
5	3 points	
6	3 points	
Total	15 points	

Chapter 1 Test

1. Solve $2392 + x = 5003$.

2. Calculate in the right order.

a. $(40 + 90) \times 2$	b. $(50 - 10) \div (5 - 3)$	c. $50 + 10 \times 4 - 20$

3. Which expression matches the problem?
   Work out the change you get if you buy seven oranges
   for $2 each, and you pay with $20.

   Also, solve the problem.

$7 \times \$2 - \$20$

$(\$20 - \$2) \times 7$

$\$20 - 7 \times \$2$

$7 \times \$20 - \$2$

4. Estimate the total cost using rounded numbers.
   Do _not_ find the exact cost.
   A game, $28.95 and TWO dolls, $14.25 each.

5. Todd had three rolls of plastic. The first one was 10 metres long,
   the second was 2 m shorter and the third was 5 m longer than the
   first one. What is the total length of the three rolls of plastic?

6. Mark the numbers and the unknown (x or ?) in the bar model.
   Write an addition or a subtraction with an unknown. Solve it.

A computer programme has been discounted by $48,
and now it costs $67. What was the original price?

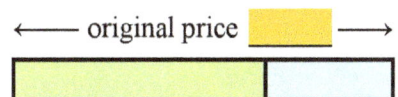

←—— original price ——→

_____

_____

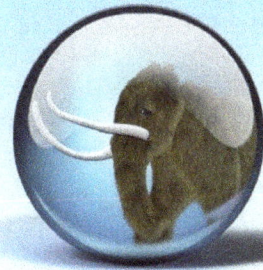

Grade 4, Chapter 2

End-of-Chapter Test

Instructions to the student:

Do not use a calculator. Answer each question in the space provided.

Instructions to the teacher:

My suggestion for grading the chapter 2 test is below. The total is 17 points. Divide the student's score by the total of 17 to get a decimal number, and change that decimal to percent to get the student's percentage score.

Question #	Max. points	Student score
1	3 points	
2	2 points	
3	3 points	
4	2 points	

Question #	Max. points	Student score
5	2 points	
6	3 points	
7	2 points	
Total	17 points	

Chapter 2 Test

1. Write the numbers.

a. 400 thousand 40	b. 4 thousand 5 hundred 60 thousand	c. 200 thousand 7 ones 6 tens

2. What is the value of the digit 8 in the following numbers?

  a. 2<u>8</u>0 340   **b.** 294 4<u>8</u>7

3. Round the numbers to the underlined place (digit).

a. 516 <u>7</u>64 ≈	**b.** 293 <u>4</u>77 ≈	**c.** 1<u>9</u>6 045 ≈

4. Calculate 225 390 − 17 692.

5. Write the numbers in order from the smallest to the greatest.

  39 294   3294   93 294 39 244 399 295

6. The king of Sookiland has 24 000 gold coins in each of his
three treasuries, plus an additional 1 382 coins in a chest.
The king of Nootyland owns a total of 78 600 gold coins.
Which king has more coins? How many more?

7. A charitable organization has one million dollars
from which they are giving $1000-grants to
students. How many students can receive the
grant?

Grade 4, Chapter 3

End-of-Chapter Test

Instructions to the student:

Do not use a calculator. Answer each question in the space provided.

Instructions to the teacher:

My suggestion for grading the chapter 3 test is below. The total is 32 points. Divide the student's score by the total of 32 to get a decimal number, and change that decimal to percent to get the student's percentage score.

Question #	Max. points	Student score
1	3 points	
2	2 points	
3	3 points	
4	3 points	
5	4 points	
6	4 points	

Question #	Max. points	Student score
7	2 points	
8a	2 points	
8b	3 points	
8c	3 points	
8d	3 points	
Total	32 points	

Chapter 3 Test

1. Multiply in your head.

   a. $4 \times 18 =$ _____

   b. $7 \times 26 =$ _____

   c. $3 \times 709 =$ _____

2. To hire a tutor for one month costs $419.85.
   Estimate how much it would cost to hire a tutor for seven months.
   (Use a rounded number.)

3. Multiply.

   a. $700 \times 9 =$ _____

   b. $120 \times 50 =$ _____

   c. $800 \times 200 =$ _____

4. Find the missing factor.

a. _____ $\times 20 = 4000$	**b.** $90 \times$ _____ $= 18\,000$	**c.** _____ $\times 700 = 5600$

5. Solve mentally.

a. $(1500 + 1500) \times 6 =$	**b.** $80 \times (1000 - 400) =$
c. $(34 + 29) \times 0 + 1293 =$	**d.** $(40 - 20) \times 4 + 40 \times 50 =$

6. Multiply.

   a.
   $$\begin{array}{r} 1\,5 \\ \times\ 7\,8 \\ \hline \end{array}$$

   b.
   $$\begin{array}{r} 7\,3\,1 \\ \times\ \ \ \ 8 \\ \hline \end{array}$$

   c.
   $$\begin{array}{r} 5\,5 \\ \times\ 1\,9 \\ \hline \end{array}$$

   d.
   $$\begin{array}{r} 5\,3\,0\,8 \\ \times\ \ \ \ \ 3 \\ \hline \end{array}$$

7. Calculate $2 \times (48 - 8) \times 17$.

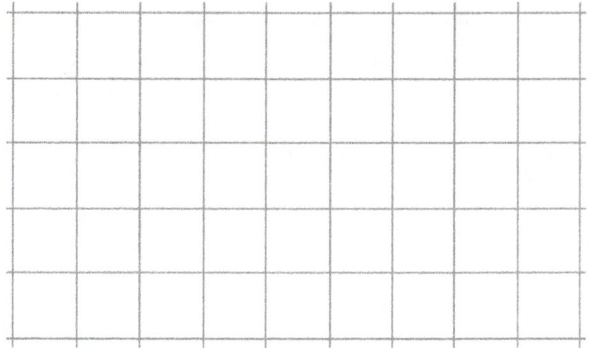

8. Solve.

a. Four folders cost $12.
How much would seven folders cost?

b. Charlotte has $30. How much money will
she have after she buys seven baskets for
$2.55 each?

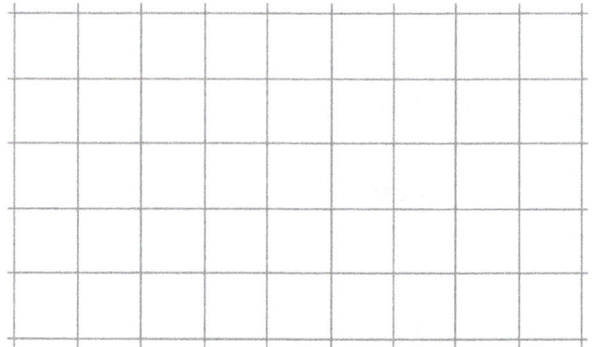

c. Lisa went shopping and bought three pairs of
jeans for $12.55 each, and a shirt for $8.90.
Now she has $13.45 left. How much money
did she take with her when she went shopping?

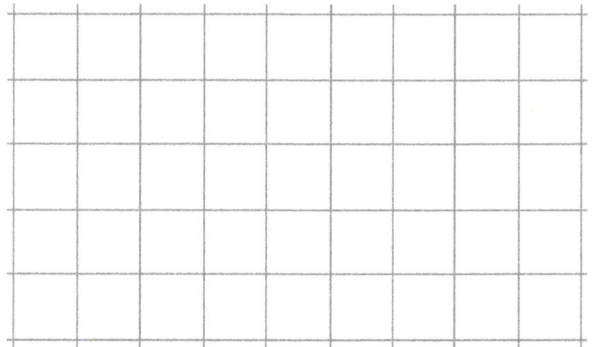

d. Isaac bought five garden hoses that were discounted. He paid $100
for them. Without the discount, he would have paid $50 more.
How much did one garden hose cost before the discount?

math
MAMMOTH

Grade 4, Chapter 4
End-of-Chapter Test

Instructions to the student:

Do not use a calculator. Answer each question in the space provided.

Instructions to the teacher:

My suggestion for grading the chapter 4 test is below. The total is 26 points. Divide the student's score by the total of 26 to get a decimal number, and change that decimal to percent to get the student's percentage score.

Question #	Max. points	Student score
1	3 points	
2	2 points	
3	9 points	
4	2 points	

Question #	Max. points	Student score
5	4 points	
6	4 points	
7	2 points	
Total	26 points	

Chapter 4 Test

1. Dinner needs to be ready by 6 p.m. and Mum needs 40 minutes to fix it. She also plans to visit the library for 1 1/2 hours before starting dinner. She can drive to the library in 10 minutes. What is the latest time she should go to the library in order to have enough time to go there and to come home and fix dinner?

2. Measure the lines below in centimetres and millimetres.

   a. _____ cm _____ mm

   b. _____ cm _____ mm

3. Convert between the different measuring units.

a.	b.
4 m 2 cm = _____ cm	2 L 80 ml = _____ ml
76 cm = _____ mm	7 m 5 cm = _____ cm
5000 mm = _____ m	4 kg 500 g = _____ g

4. What is the perimeter of a square with 2 cm 6 mm sides?

5. You can buy liquid dish soap in 1000-ml bottles or in 400-ml bottles.

   a. How many of the bigger bottles do you need to buy to get 2 litres of dish soap?

   b. How many of the smaller bottles do you need to buy to get 2 litres of dish soap?

6. The dish soap in 1000-ml bottles costs $13.80 per bottle.
   The dish soap in 400-ml bottles costs $12.10 per bottle.

   a. How much does it cost to purchase 2 litres of the first kind of dish soap?

   b. How much does it cost to purchase 2 litres of the second kind of dish soap?

7. Mrs. Banks put 1 litre of honey into 125 ml jars. How many jars did she fill?

Grade 4, Chapter 5

End-of-Chapter Test

Instructions to the student:

Do not use a calculator. Answer each question in the space provided.

Instructions to the teacher:

My suggestion for grading the chapter 5 test is below. The total is 37 points. Divide the student's score by the total of 37 to get a decimal number, and change that decimal to percent to get the student's percentage score.

Question #	Max. points	Student score
1	6 points	
2	2 points	
3	2 points	
4	3 points	
5	4 points	
6	8 points	

Question #	Max. points	Student score
7	3 points	
8	3 points	
9	3 points	
10	3 points	
Total	37 points	

Chapter 5 Test

1. Solve.

a.	b.	c.
$13 \div 4 = $ _____ R _____	$33 \div 7 = $ _____ R _____	$40 \div 12 = $ _____ R _____
$13 \div 5 = $ _____ R _____	$45 \div 8 = $ _____ R _____	$67 \div 9 = $ _____ R _____

2. If seven metres of fabric cost $84, how much would five metres of fabric cost?

3. Seth had saved $350. He spent $\frac{3}{5}$ of that to buy a camera.

   How much did the camera cost?

4. Paul sold 2/3 of his 1200 bricks to his neighbour.
   Then, Paul sold another 150 bricks.
   Now how many bricks does he have left?

5. Solve. Check by multiplying.

a. $565 \div 5$ Check:	**b.** $3664 \div 8$ Check:

6. Find all the factors of the given numbers.

a. 28 factors:	b. 13 factors:
c. 32 factors:	d. 76 factors:

7. Travis shared 125 pencils between seven children, as equally as he could.

How many pencils did each child get?

How many pencils were left over?

8. Susan found four different pairs of dress shoes in a shop, with prices of $39, $45, $63 and $41. What is their average price?

9. Is 924 divisible by 7? Explain why or why not.

10. Place the numbers 10, 20 and 30 *and* <u>brackets</u> in the expression below so that the answer is more than three hundred. Do not forget the brackets!

_____ − _____ × _____ = _____

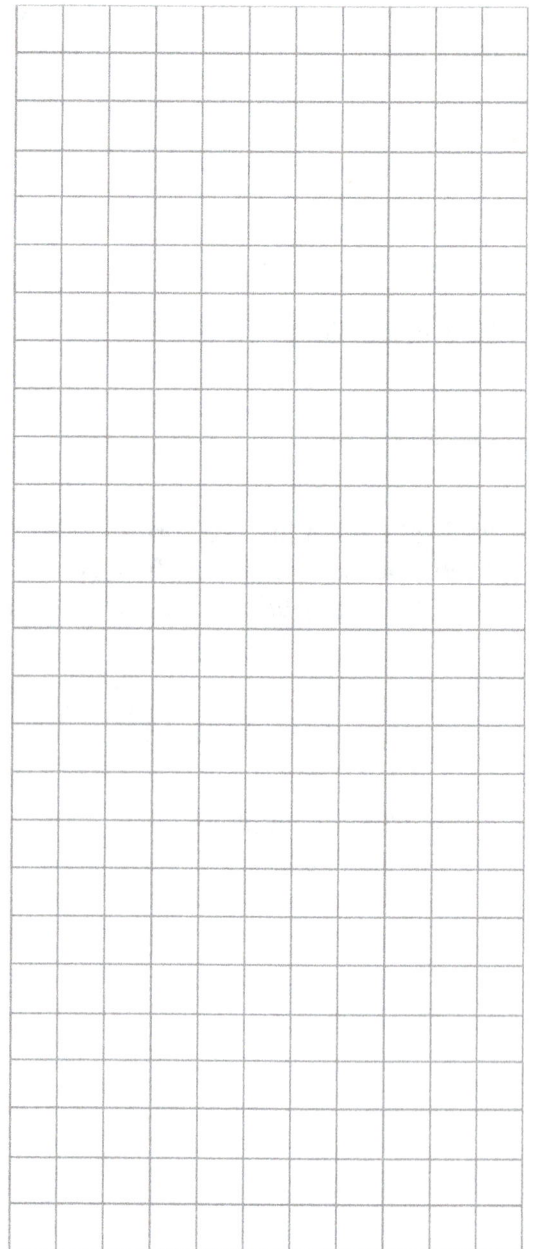

Grade 4, Chapter 6

End-of-Chapter Test

Instructions to the student:

Do not use a calculator. Answer each question in the space provided.

Instructions to the teacher:

My suggestion for grading the chapter 6 test is below. The total is 22 points. Divide the student's score by the total of 22 to get a decimal number, and change that decimal to percent to get the student's percentage score.

Question #	Max. points	Student score
1	2 points	
2	2 points	
3	2 points	
4	5 points	
5	2 points	

Question #	Max. points	Student score
6	4 points	
7	2 points	
8	3 points	
Total	22 points	

Chapter 6 Test

1. Draw a 75° angle.

2. Measure this angle.

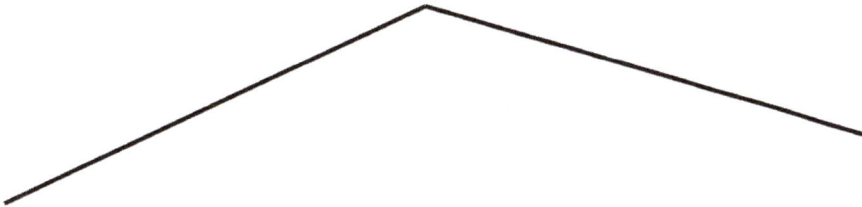

3. What is the angle measure of the angle x?

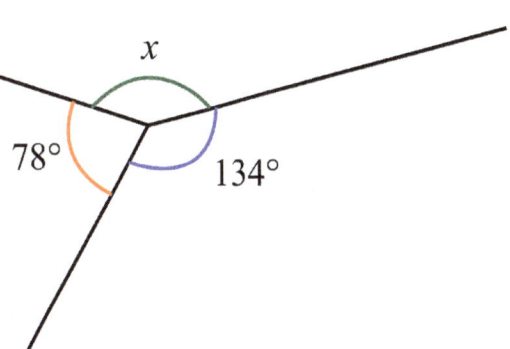

    78° x 134°

4. **a.** What is this shape called?

    b. Draw a diagonal into it (a line from one corner to another) so that you will get *two obtuse* triangles.

    c. Measure the perimeter of one of the obtuse triangles in centimetres and millimetres.

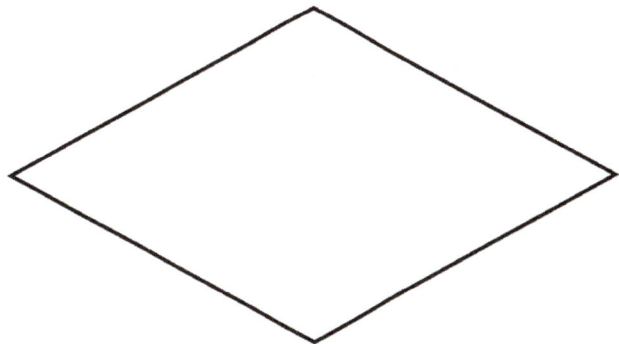

5. Draw a quadrilateral that has ONE
   right angle, and the other angles
   are not right angles.

6. **a.** Draw a right triangle.

   b. Measure all the angles of your triangle.

7. Classify these triangles according to their angles.   **a.**   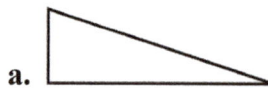   **b.**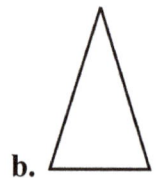

8. Find the area of the *coloured* area
   in square metres.

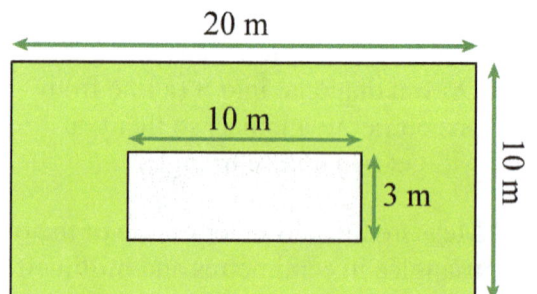

20 m

10 m

3 m

10 m

Grade 4, Chapter 7

End-of-Chapter Test

Instructions to the student:

Do not use a calculator. Answer each question in the space provided.

Instructions to the teacher:

My suggestion for grading the chapter 7 test is below. The total is 25 points. Divide the student's score by the total of 25 to get a decimal number, and change that decimal to percent to get the student's percentage score.

Question #	Max. points	Student score
1	6 points	
2	6 points	
3	2 points	
4	4 points	

Question #	Max. points	Student score
5	3 points	
6	4 points	
Total	25 points	

Chapter 7 Test

1. Add and subtract.

a. $\dfrac{3}{6} + \dfrac{2}{6} + \dfrac{1}{6} =$	**b.** $1\dfrac{2}{3} + \dfrac{2}{3} =$	**c.** $3\dfrac{3}{5} + 2\dfrac{1}{5} =$
d. $\dfrac{11}{12} - \dfrac{7}{12} - \dfrac{2}{12} =$	**e.** $3\dfrac{1}{5} - \dfrac{3}{5} =$	**f.** $7\dfrac{5}{6} - 2\dfrac{1}{6} =$

2. Arrange the fractions in order from smallest to greatest.

a. $\dfrac{3}{4} , \dfrac{3}{8} , \dfrac{1}{2}$	**b.** $\dfrac{5}{5} , \dfrac{5}{7} , \dfrac{7}{5}$	**c.** $\dfrac{5}{9} , \dfrac{5}{2} , \dfrac{5}{6}$

3. Split both the coloured and white pieces as asked. Write the fraction before and the fraction after.

a. Split all pieces into four new ones.

b. Split all pieces into three new ones.

4. Write the equivalent fraction. Use multiplication.

a. $\dfrac{1}{5} = \dfrac{2}{\boxed{}}$	**b.** $\dfrac{3}{4} = \dfrac{\boxed{}}{12}$	**c.** $\dfrac{4}{5} = \dfrac{\boxed{}}{25}$	**d.** $\dfrac{1}{6} = \dfrac{4}{\boxed{}}$

5. Multiply. Give your answer as a whole number or as a mixed number.

a. $3 \times \dfrac{4}{10} =$	**b.** $5 \times \dfrac{3}{5} =$	**c.** $\dfrac{3}{8} \times 4 =$

6. Three brothers made a big pizza and divided it into 12 equal pieces. Patrick ate 1/4 of the pizza, Peyton ate 1/12 of it, and Louis ate three times as much as Peyton.

a. Who ate the most pizza?

b. How much more did Louis eat than Peyton?

Grade 4, Chapter 8

End-of-Chapter Test

Instructions to the student:

Do not use a calculator. Answer each question in the space provided.

Instructions to the teacher:

My suggestion for grading the chapter 8 test is below. The total is 28 points. Divide the student's score by the total of 28 to get a decimal number, and change that decimal to percent to get the student's percentage score.

Question #	Max. points	Student score
1	4 points	
2	5 points	
3	6 points	
4	5 points	

Question #	Max. points	Student score
5	2 points	
6	2 points	
7	4 points	
Total	28 points	

Chapter 8 Test

1. Mark these decimals on the number line: 1.60   1.21   1.78   1.04

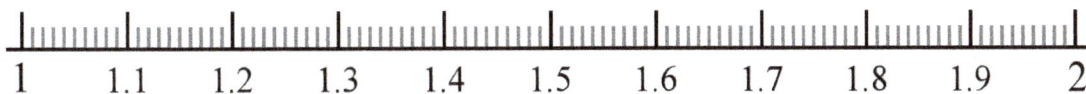

```
|ıııııııııı|ıııııııııı|ıııııııııı|ıııııııııı|ıııııııııı|ıııııııııı|ıııııııııı|ıııııııııı|ıııııııııı|ıııııııııı|
1       1.1      1.2      1.3      1.4      1.5      1.6      1.7      1.8      1.9       2
```

2. Write the fractions as decimals and decimals as fractions.

a. $\dfrac{2}{10}$	b. $7\dfrac{4}{100}$	c. $\dfrac{74}{100}$	d. 0.52	e. 3.9

3. Add and subtract.

a. $0.5 + 1.7 = $ _____   b. $0.44 + 0.51 = $ _____   c. $0.2 - 0.01 = $ _____

d. $1.6 - 0.9 = $ _____   e. $0.3 + 0.07 = $ _____   f. $5.05 - 2.01 = $ _____

4. Compare. Write $<$, $>$ or $=$ between the numbers.

a. 0.4 ☐ 0.14	b. 2.9 ☐ 2.90	c. 4.3 ☐ 4.03	d. 0.45 ☐ $\dfrac{1}{2}$	f. 7.18 ☐ 7.8

5. Write in order from the smallest to the greatest number: 7.2   2.7   2.07   2.17   2.77

6. Find the total weight of four books that weigh 1.3 kg each.

7. Calculate.

a. $4.56 + 2.8$	b. $4.56 - 2.8$

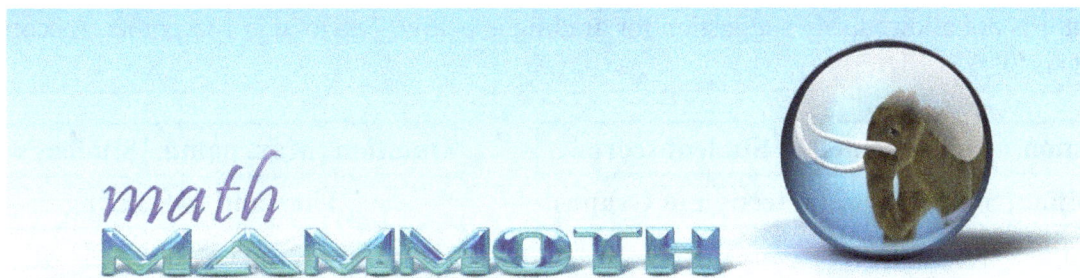

End-of-the-Year Test Grade 4
International Version

This test is quite long, so I don't recommend that you have your child/student do it in one sitting. Break it into parts and administer them over several days. Use your judgement.

This test works as a diagnostic test. So, you may even skip those areas and concepts that you already know for sure your student has mastered.

The test does not cover every single concept that is covered in the *Math Mammoth Grade 4 Complete Curriculum,* but all of the major concepts and ideas are tested here. This test is evaluating the student's ability in the following content areas:

- addition and subtraction
- early algebraic thinking
- the order of operations
- graphs
- large numbers and place value
- rounding and estimating
- multi-digit multiplication
- word problems
- some basic conversions between measuring units
- measuring length
- time calculations
- long division
- the concept of remainder
- factors
- area and perimeter
- measuring and drawing angles
- classifying triangles according to their angles
- adding and subtracting fractions and mixed numbers (like fractional parts)
- equivalent fractions
- comparing fractions
- multiplying fractions by whole numbers
- the concept of a decimal (tenths/hundredths)
- comparing decimals

In order to continue with *Math Mammoth Grade 5 Complete Curriculum*, I recommend that the student gain a **minimum score of 80%** on this test, and that the teacher or parent revise with him any content areas in which the student is weak. Students scoring between 70% and 80% may also go on to grade 5, depending on the types of errors (careless errors or not remembering something, versus the lack of understanding). The most important content areas to master are multi-digit multiplication, long division, place value and word problems. Again, use your judgement.

A calculator is not allowed. My suggestion for grading is below. The total is 185 points. A score of 148 points is 80%.

Question	Max. points	Student score
Addition, Subtraction, Patterns and Graphs		
1	2 points	
2a	1 point	
2b	2 points	
3	2 points	
4	6 points	
5	4 points	
6	2 points	
7	4 points	
8	3 points	
subtotal		/ 26
Large Numbers and Place Value		
9	3 points	
10	2 points	
11	3 points	
12	3 points	
13	2 points	
14	3 points	
15	3 points	
16	4 points	
subtotal		/ 23
Multi-Digit Multiplication		
17	6 points	
18	3 points	
19	8 points	
20	3 points	
21a	3 points	
21b	2 points	
21c	2 points	
21d	3 points	
subtotal		/ 30

Question	Max. points	Student score
Time and Measuring		
22	2 points	
23	1 point	
24	3 points	
25	2 points	
26	6 points	
27	2 points	
28	1 point	
29	2 points	
subtotal		/ 19
Division and Factors		
30	4 points	
31	3 points	
32	4 points	
33a	3 points	
33b	2 points	
34	6 points	
35	4 points	
36	2 points	
37	4 points	
subtotal		/ 32
Geometry		
38	2 points	
39	2 points	
40	3 points	
41	2 points	
42	2 points	
43	1 point	
44	2 points	
45	3 points	
subtotal		/ 17

Question	Max. points	Student score
Fractions and Decimals		
46	1 point	
47	1 point	
48	3 points	
49	2 points	
50	4 points	
51	4 points	
52	2 points	
53	3 points	
54	4 points	
55	4 points	
56	4 points	
57	4 points	
58	2 points	
	subtotal	/ 38
	TOTAL	/ 185

End of the Year Test - Grade 4

Addition, Subtraction, Patterns and Graphs

1. Subtract. Check by adding.

$5200 - 2677 - 543$	Add to check:

2. **a.** Round the prices to the nearest dollar. Use the rounded prices to estimate the total cost.

   Crackers $2.25; cheese $8.90; jam $4.75; butter $9.30.

   b. Now, use the exact prices (not rounded prices). Mrs. Grayson bought the items listed above and paid with $50. How much was her change?

3. *Estimate* the cost of buying five notebooks for $2.85 each and two pencil cases for $3.25 each.

4. Calculate in the right order.

a. $3 \times (4 + 6) =$ _____	**b.** $3 \times 3 + 8 \div 4 =$ _____	**c.** $20 \times 3 + 80 \div 1 =$ _____
$100 - 4 \times 4 =$ _____	$(7 - 3) \times 3 + 2 =$ _____	$15 + 2 \times (8 - 6) =$ _____

5. Circle the number sentence that fits the problem. Then solve for x.

a. Alicia had $35. Then she earned some money (x). Now she has $92. $35 + x = 92$ OR $35 + 92 = x$ $x =$ _____	**b.** Mike baked cookies and gave 24 of them to a friend and now he has 37 cookies left. $37 - 24 = x$ OR $x - 24 = 37$ $x =$ _____

6. **a.** Continue this pattern for four more numbers:

   2 000   1 750   1 500   1 250

   b. Write a list of six numbers that follows this pattern: Start at 200, and add 300 each time.

7. These are the quiz scores for several students. 2 5 8 7 6 6 7 10 10 4 7 7 8 6 8 5 9 9 8 6 6 5 7 9
Make a frequency table and a bar graph.

Quiz score	Frequency
1	
2	
3	
4	
5	
6	
7	
8	
9	
10	

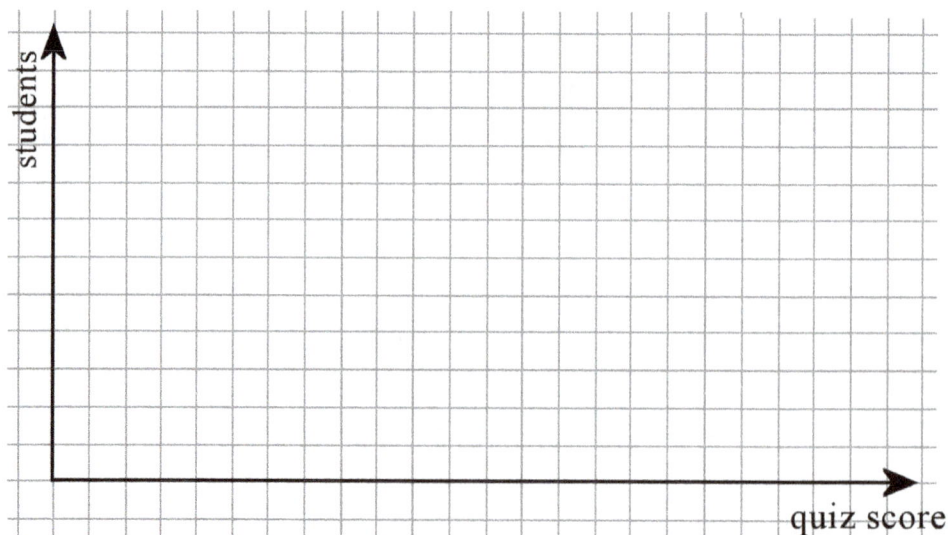

8. Write an addition or a subtraction with an unknown (*x* or ?). Solve it. The bar model can help.

A doll used to cost $27.95 but now the price is $21.45. How much is the discount?

←—— original price ——→

_____

_____

Large Numbers and Place Value

9. Subtract from whole thousands.

a. 2000 − 1 = _____	**b.** 5000 − 20 = _____	**c.** 6000 − 300 = _____

10. Write the numbers in normal form.

   a. 800 thousand 50

   b. 25 thousand 4 hundred 7

11. Find the missing numbers.

a. 30 550 = 50 + _____ + 500	**b.** 809 100 = 800 000 + 100 + _____
c. 725 608 = 20 000 + 700 000 + 8 + _____ + 5000	

12. Compare, writing <, > or = between the numbers.

a. 54 500 55 400	**b.** 108 882 108 828	**c.** 71 600 61 700

13. Write the numbers in order from the smallest to the greatest:

   217 200   227 712   27 200   227 200

14. Round the numbers as the dashed line indicates (to the underlined digit).

a. 43̲6 102 ≈	**b.** 8̲9 756 ≈	**c.** 27 5̲29 ≈

15. Round to the nearest ten thousand.

a. 426 889 ≈	**b.** 495 304 ≈	**c.** 7345 ≈

16. Calculate. Line up all the digits carefully.

a. 476 708 + 24 392 + 563

b. 405 112 − 81 424

Multi-Digit Multiplication

17. Multiply, and find the missing factors.

a. $70 \times 3 =$ _____	**b.** $6 \times 800 =$ _____	**c.** $40 \times 80 =$ _____
d. _____ $\times 3 = 360$	**e.** $50 \times$ _____ $= 4000$	**f.** _____ $\times 300 = 21\,000$

18. Tom earns $20 per hour.

a. How much will he earn in an 8-hour workday? _____

b. How much will he earn in a 40-hour workweek? _____

c. How many days will he need to work in order to earn at least $600? _____

19. Multiply. Estimate the answer on the line.

a. 5×196	**b.** 35×38	**c.** 7×3188	**d.** 89×22
≈ _____	≈ _____	≈ _____	≈ _____

20. Write the area of the *whole* rectangle as a SUM of the areas of the *smaller* rectangles. Lastly, add to find the total area.

Area = 8 × 127

= ___ × _____ + ___ × ___ + ___ × ___

=

100   20 7

8

21. Solve the problems. **Write a number sentence** or several for each problem.

a. Work out the change, if Susan buys 26 books for $14 each, and pays with $400.

_____

_____

_____

b. How many minutes are there in a day (24 hours)?

_____

_____

c. One side of a square is 375 cm. What is its perimeter?

_____

_____

d. Schoolbags costing $277 are discounted by $58. Aunt Patricia buys eight for presents. What is the total cost?

_____

_____

Time and Measuring

22. Measure the lines in centimetres and millimetres.

    a. _____ cm _____ mm

    b. _____ cm _____ mm

23. How much time passes from 10:54 a.m. to 5:06 p.m.?

24. Lyle kept track of how long it took him to do
    his homework:

Monday	Tuesday	Wednesday	Thursday	Sunday
1 h 45 min	50 min	1 h 15 min	2 h 15 min	55 min

    How much time did he spend with homework in total?

    _____

25. A teacher started her workday at 7:00 am, and stopped it at 3:35 pm.
    But in between, she had a 45-minute lunch break, and another break
    of 20 minutes. How many hours/minutes did she actually work?

26. Convert between the different measuring units.

a.	b.	c.
2 kg = _____ g	5 L 200 ml = _____ ml	8 cm 2 mm = _____ mm
11 kg 600 g = _____ g	3 m = _____ cm	10 km = _____ m

27. George jogs daily on a track through the woods that is 3 km 800 m long.
   What is the total distance he runs in four days?

28. Bonnie drank 350 ml of a 2-litre bottle of water.
   How much is left?

29. The long sides of a rectangle measure 5 m 20 cm,
   and the short sides are 3 m 4 cm.

   What is the perimeter? _____ m _____ cm

Division and Factors

30. Divide. Check each problem by multiplying.

a. 567 ÷ 9 Check:	b. 8564 ÷ 4 Check:

31. Solve.

| a. 47 ÷ 5 = _____ R ___ | b. 25 ÷ 3 = _____ R ___ | c. 57 ÷ 9 = _____ R ___ |

32. Solve.

a. Amanda put 48 photographs into an online photo album. On each page she could fit nine photos.

How many photographs were on the last page?

How many pages were full?

b. If you buy a 15-metre roll of chain-link fence that costs $255, and then you sell 3 metres of it to your neighbour, how much should your neighbour pay?

33. Solve.

a. Mitch had saved $264. He spent 3/8 of it to buy a book. How much did the book cost?

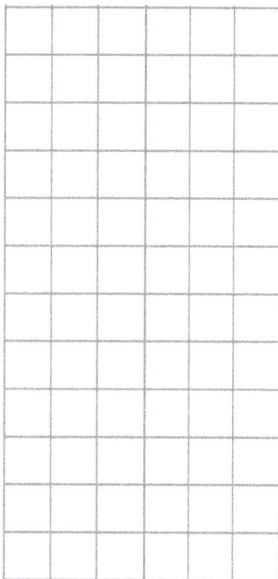

b. Mary packed 117 muffins into bags of six. How many bags does Mary need for them?

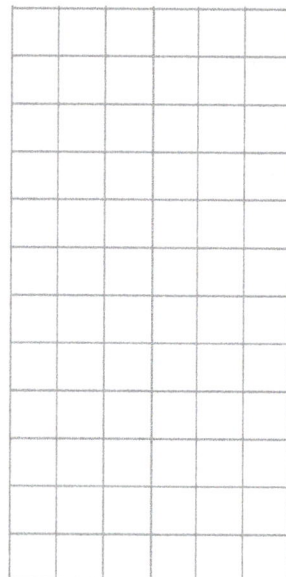

34. Mark with an X if the number is divisible by the given numbers.

number	divisible by 1	divisible by 2	divisible by 3	divisible by 4	divisible by 5	divisible by 6	divisible by 7	divisible by 8	divisible by 9	divisible by 10
80										
75										
47										

35. Fill in.

a. Is 5 a factor of 60? _____, because _____ × _____ = _____ .	**b.** Is 7 a divisor of 43? _____, because _____ ÷ _____ = _____ .
c. Is 96 divisible by 4? _____, because _____ .	**d.** Is 34 a multiple of 7? _____, because _____ .

36. List three prime numbers.

37. Find all the factors of the given numbers.

a. 56 factors:	**b.** 78 factors:

Geometry

38. Measure this angle.

39. Draw below an angle of 65°.

40. Draw below any obtuse triangle,
    and measure its angles.

41. Write an addition sentence about the
    angle measures. Use an unknown (x)
    for one angle measure.

    Then solve it.

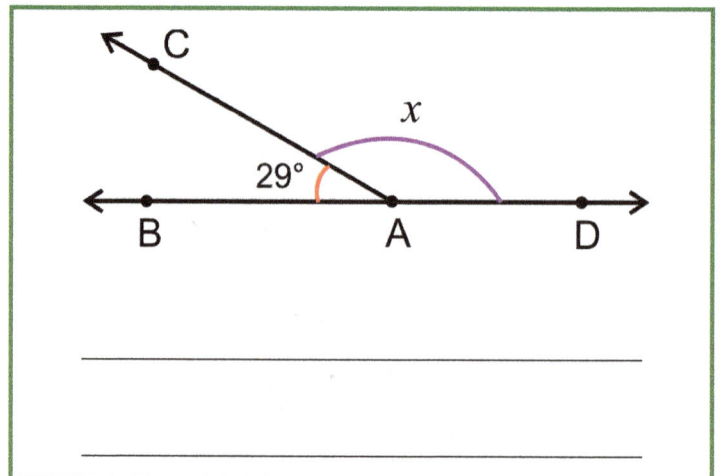

_____

_____

42. Sketch here any rectangle. Then draw one diagonal
    line in it (a line from corner to corner).
    What kind of triangles are formed?

43. Draw two line segments that are
    perpendicular to each other.

44. Draw as many different
    symmetry lines as you
    can into this shape.

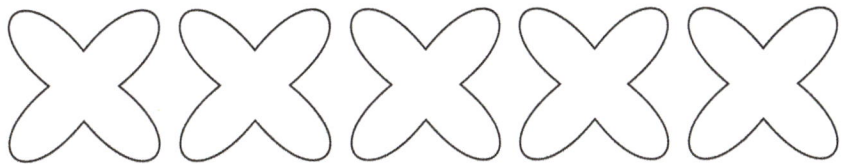

45. This picture shows the floor of a room with a carpet on the floor.
    The room itself measures 9 metres by 4 metres. The carpet is 2
    metres by 3 metres. Find the area of floor outside the carpet (not
    including the carpet).

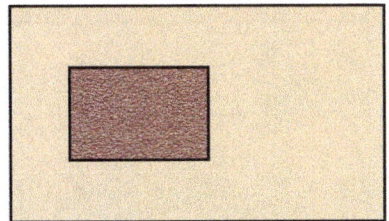

Fractions and Decimals

46. Write an addition to match the picture:

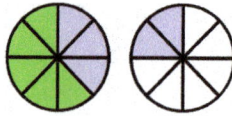

47. Emma did 1/4 of a puzzle, and
    Mum did another fourth of it.
    How much of the puzzle is still left to do?

48. Add and subtract. Give your final answer as a whole number or as a mixed number if possible.

a. $\dfrac{4}{5} + \dfrac{3}{5} =$	**b.** $1\dfrac{1}{6} - \dfrac{2}{6} =$	**c.** $3\dfrac{6}{8} + 2\dfrac{2}{8} =$

49. Split the existing pieces.
    Fill in the missing parts.

a. Each piece is split into 2 new ones.

$$\frac{4}{5} = \frac{}{}$$

b. Each piece is split into ____ new ones.

$$\frac{}{} = \frac{6}{9}$$

50. Write the equivalent fractions.

a. $\dfrac{2}{3} = \dfrac{}{15}$	**b.** $\dfrac{3}{5} = \dfrac{9}{}$	**c.** $\dfrac{1}{6} = \dfrac{}{12}$	**d.** $\dfrac{1}{3} = \dfrac{}{9}$

51. Compare the fractions.

a. $\dfrac{2}{3} \square \dfrac{3}{8}$ **b.** $\dfrac{6}{5} \square \dfrac{7}{8}$ **c.** $\dfrac{11}{12} \square \dfrac{11}{10}$ **d.** $\dfrac{1}{3} \square \dfrac{5}{12}$

52. Write these fractions in order, from the smallest to the greatest: $\dfrac{5}{4}, \dfrac{7}{10}, \dfrac{65}{100}$

53. Fill in.

a. $\dfrac{3}{8} = 3 \times \underline{\hphantom{xxx}}$	b. $4 \times \dfrac{2}{5} =$	c. $7 \times \dfrac{2}{12} =$

54. Mark on the number line the following decimals: 0.55 0.08 0.27 0.80

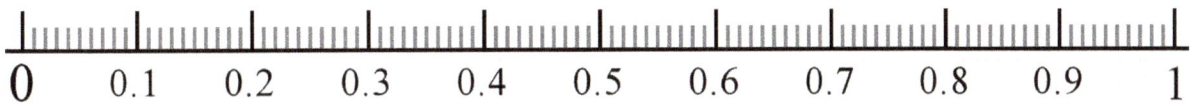

```
|........|........|........|........|........|........|........|........|........|........|
0     0.1     0.2     0.3     0.4     0.5     0.6     0.7     0.8     0.9      1
```

55. Write the fractions and mixed numbers as decimals.

a. $\dfrac{3}{10}$	b. $3\dfrac{9}{10}$	c. $\dfrac{9}{100}$	d. $7\dfrac{45}{100}$

56. Write the decimals as fractions or mixed numbers.

a. 0.6	b. 6.7	c. 0.21	d. 5.05

57. Compare.

   a. 0.17 ☐ 0.2   b. 1.6 ☐ 1.56   c. 13.09 ☐ 13.9   d. 9.80 ☐ 9.8

58. Add and subtract.

a. $7.81 + 5.2$	b. $6.1 - 2.36$

Using the cumulative revisions and the worksheet maker

The cumulative revisions contain a mix of problems, practising topics from the curriculum up to the chapter named in the revision. For example, the cumulative revision for chapters 1-6 may include problems matching chapters 1, 2, 3, 4, 5, and 6. It can be used any time after the student has studied the curriculum through chapter 6.

These cumulative revision lessons are optional; use them as needed. The student doesn't have to complete all the cumulative revisions, however I recommend using at least three of these revisions during the school year. The teacher can also use the revisions as diagnostic tests to find out what topics the student has trouble with.

Math Mammoth complete curriculum also includes an easy worksheet maker, which is the perfect tool to make more problems for children who need more practice. The worksheet maker covers most topics in the curriculum, excluding word problems. Most people find it to be a very helpful addition to the curriculum.

The download version of the curriculum includes the worksheet maker as a file, and you can also access the worksheet maker online at

https://www.mathmammoth.com/private/Make_extra_worksheets_grade4.htm

In addition to the cumulative revision and the worksheet maker, we also offer a free online practice area at https://www.mathmammoth.com/practice/. This section of the website has a growing number of games and practice activities for many maths topics.

Cumulative Revision, Grade 4, Chapters 1-2

1. Add mentally. You can add in parts (tens and ones separately), or use other "tricks."

a. $56 + 82 =$ _____	**b.** $29 + 29 =$ _____	**c.** $69 + 58 =$ _____
$27 + 47 =$ _____	$34 + 58 =$ _____	$25 + 45 =$ _____
$22 + 81 =$ _____	$99 + 45 =$ _____	$72 + 72 =$ _____

2. Solve in the correct order.

a. $(400 + 200) \times 3 =$ _____	**b.** $10 \times (50 + 10) =$ _____
$400 + 200 \times 3 =$ _____	$10 \times 50 + 10 =$ _____
c. $6 + 9 \div 3 =$ _____	**d.** $8 \times (300 - 200) - 300 =$ _____
$80 \div 20 \div 4 =$ _____	$(70 - 30) \times 4 - 20 =$ _____

3. **a.** Continue this pattern: subtract _____ each time.

700	620	540					

b. Continue this pattern: add 99 each time, starting at 0

0							

4. Write an addition with an unknown (x). Mark the numbers and the unknown in the bar model. Solve.

A shipment of toy cars contained 1000 cars. Of them, 450 were SUVs, 128 were vans, and the rest were regular cars. How many regular cars were there?

Addition:

Solution: $x =$ _____

5. Add in columns.

a.
```
    1 9 1
  2 0 3 5
    8 7 3
  1 0 1 8
    3 0 1
+     2 7
_____
```

b.
```
    4 3 5
    3 4 9
      2 0
  1 8 1 1
    2 9 4
  9 4 9 3
+   9 7 6
_____
```

c.
```
  3 3 7 9 0
  2 3 1 7 6
    7 4 6 3
6 5 1 0 0 6
      5 1 7
+       9 9
_____
```

6. Write the numbers in order from the smallest to the greatest.

   a. 18 399   819 090   8030   818 939

   b. 52 200   5220   250 500   520 500

7. Write the numbers.

   a. 284 thousand 1

			,			

   b. 50 thousand 50

			,			

8. What is the _value_ of the underlined digit in the following numbers?

   a. 212 047

   b. 94 012

   c. 500 049

   d. 249 255

9. Round the numbers to the nearest hundred.

a. 7520 ≈ _____	b. 2712 ≈ _____	c. 3953 ≈ _____
d. 354 ≈ _____	e. 56 278 ≈ _____	f. 293 596 ≈ _____

10. A tablet device was discounted twice: first by $30, then by another $25.
    Now it costs $176. What was the original price?

58

Cumulative Revision, Grade 4, Chapters 1-3

1. Subtract and compare the results. The problems are "related" – can you see how?

a. 15 − 6 = _____	**b.** 14 − 8 = _____	**c.** 12 − 7 = _____
65 − 6 = _____	74 − 8 = _____	82 − 7 = _____
650 − 60 = _____	240 − 80 = _____	1200 − 700 = _____
250 − 60 = _____	1400 − 800 = _____	620 − 70 = _____

2. The table lists the sales that Jenny had for selling homemade dresses.
   Find her total sales over these five weeks.

Week 37	Week 38	Week 39	Week 40	Week 41
$458	$366	$427	$503	$413

3. Fill in.

   a. A car travels at 80 kilometres per hour.

Kilometres								
Hours	1	5	7	9	10	12	15	20

   b. George bought fencing for his dog kennel. Four metres cost $36.

Dollars								
Metres	1	2	3	4	5	8	10	15

4. Write the numbers in order.

   a. 5500   5005   5604   5000   1554

   _____ < _____ < _____ < _____ < _____

   b. 37 700   73 737   38 707   307 988   3800

   _____ < _____ < _____ < _____ < _____

5. Add in columns in the grid provided below.

a. $851\,091 + 40\,510 + 91\,576$

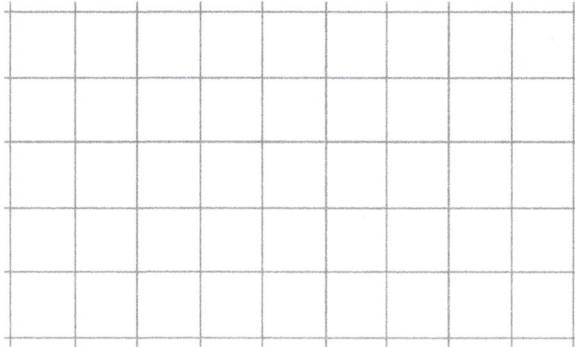

b. $39\,312 + 506\,636 + 9382$

6. Round to the nearest dollar.

a. $\$3.05 \approx$

b. $\$8.35 \approx$

c. $\$25.90 \approx$

7. Beth and Gary helped their mum with a yard sale that they ran for five days. The graph shows how much they earned each day.

a. Estimate their total earnings for these five days.

b. Estimate how much less they earned on their worst day than on their best day.

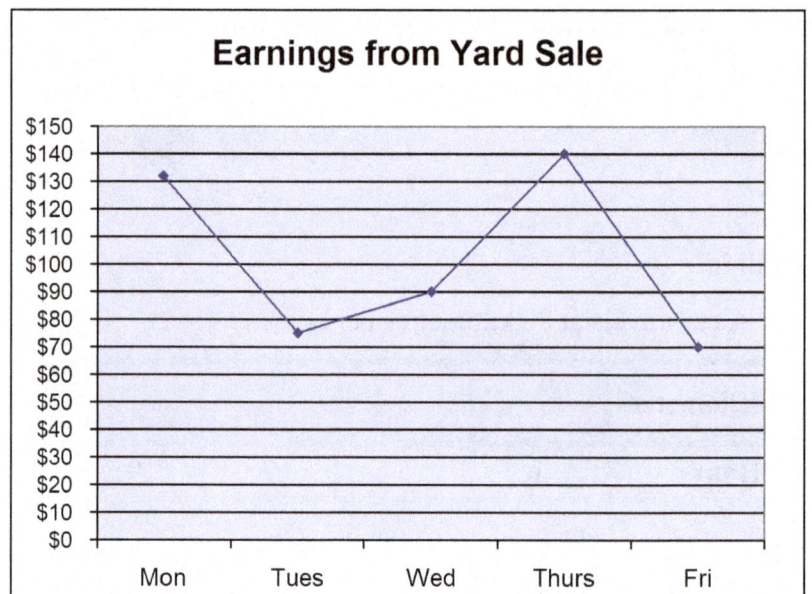

Earnings from Yard Sale

8. Jesse read the whole year's issues of Motorcycle Magazine, with 96 pages in each monthly issue. He spent about two-and-a-half hours reading each magazine.

a. How many pages did he read?

b. About how long did it take him to read *half* of the magazines?

Cumulative Revision, Grade 4, Chapters 1-4

1. Multiply.

a.
$$\begin{array}{r} 3\ 2\ 5 \\ \times\ \ \ \ \ 3 \\ \hline \end{array}$$

b.
$$\begin{array}{r} 4\ 1\ 9\ 8 \\ \times\ \ \ \ \ \ \ \ 5 \\ \hline \end{array}$$

c.
$$\begin{array}{r} 8\ 2 \\ \times\ 2\ 4 \\ \hline \end{array}$$

d.
$$\begin{array}{r} 5\ 6 \\ \times\ 7\ 3 \\ \hline \end{array}$$

2. Write an addition sentence using x. Solve.

The auditorium has 609 seats. Of those, 256 seats are reserved, and the rest are non-reserved. How many seats are non-reserved?

_____ + _____ = _____

$x =$

3. Do the calculations in the right order.

a. $3 \times (6 + 3) =$ _____	b. $(11 - 4) \times 8 + 1 =$ _____
c. $24 - 4 \div 2 =$ _____	d. $60 - 1 \times 7 + 12 \div 3 =$ _____
e. $(22 - 16) \times 3 + 3 =$ _____	f. $72 \div (6 + 6) - 5 =$ _____

4. Round these numbers to the nearest hundred.

a. $555 \approx$ _____	b. $8889 \approx$ _____	c. $351931 \approx$ _____
d. $64 \approx$ _____	e. $244295 \approx$ _____	f. $38009 \approx$ _____

5. Write the numbers.

    a. three hundred and five thousand two hundred

    b. forty thousand and thirty-three

6. Draw a line in the numbers to separate the thousands. Compare. Write either < or > in between the numbers.

a. 7 2 3 0 5 0 6 9 9 0 9 9		**b.** 3 2 2 3 2 0 3 2 2 3 2 2	
c. 6 9 2 1 5 9 6 9 2 1 9 6		**d.** 1 4 0 0 0 0 1 4 1 0 0	
e. 1 1 3 9 9 9 1 1 5 3 9 9		**f.** 8 3 6 4 9 6 8 8 4 8 2	

7. Multiply.

a. $100 \times 11 = $ _____	**b.** $18 \times 10 = $ _____	**c.** $100 \times 920 = $ _____
$19 \times 10 = $ _____	$4000 \times 200 = $ _____	$32 \times 2000 = $ _____
$3000 \times 40 = $ _____	$88 \times 100 = $ _____	$400 \times 22 = $ _____

8. Multiply the money amounts in parts. Give your answers in dollars.

a. 6×30c = _____ c = $ _____	**b.** 5×85c = _____ c + _____ c = $ _____
c. $6 \times \$1.70$ = _____ + _____ = $ _____	**d.** $3 \times \$2.80$ = _____ + _____ = $ _____

9. Solve the word problems.

    a. Mary planted 22 rows of corn with 38 plants in each row. *Approximately* how many corn plants are there?

    b. Jim had his car fixed. He had to pay for seven hours of work, $8.20 per hour. He paid with $100. How much was his change?

        _____

        _____

Cumulative Revision, Grade 4, Chapters 1-5

1. **a.** The beach is 1200 metres long and 1/6 of that is for boat access. How many metres of the beach are not accessible by boat?

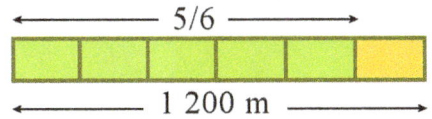

5/6

1 200 m

   b. Two-thirds of a group of students are girls. There are 11 boys. How many girls are there?

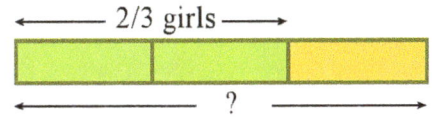

   What is the total number of boys and girls?

2/3 girls

?

2. A baker charted how much he spent on flour from January through May.

   Use rounded numbers, and estimate:

   a. *About* how much more did he spend in May than March?

   b. *About* how much did he spend in March, April and May in total?

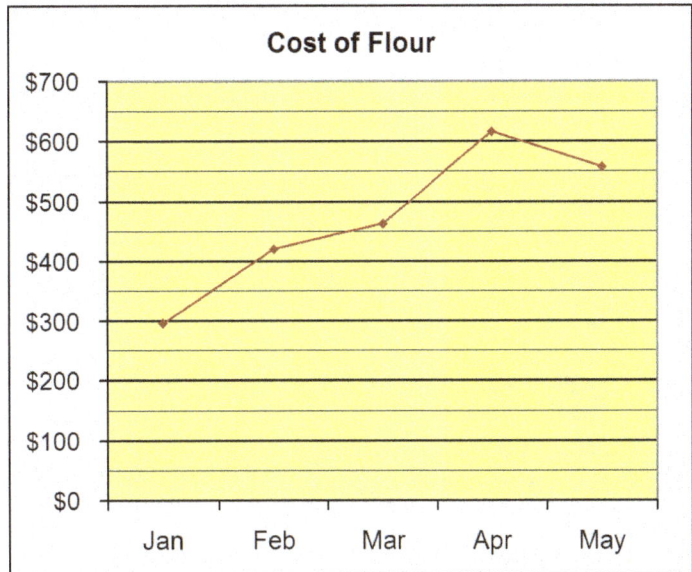

Cost of Flour

3. Solve.

   a. One triangle weighs _____.

   b. One square weighs _____.

4. Convert.

   a. 1 kg 300 g = _____ g

   4 kg 20 g = _____ g

   b. 7500 g = _____ kg _____ g

   6 kg 45 g = _____ g

5. Fill in the tables.

Minutes	1	5	6	7	10
Seconds					

Days	1	3	6	10
Hours				

6. Explain in which order to do the operations in the problem on the right. $(5 + 39) \div 4 - 2 \times 2$

First do _____, which equals _____ . Then, _____ that answer

by _____ .   This leaves _____ . Then, do _____ = _____ .

Lastly, _____ that from _____ . The answer is _____ .

7. How much time passes? You can use subtraction.

a. From 2:42 p.m. to 7:36 p.m.	**b.** From 3:39 p.m. to 11:03 p.m.	**c.** From 8:45 to 17:09.
h m − h m _____	h m − h m _____	h m − h m _____

8. Convert between the measures of volume.

a. 6 L = _____ ml	**b.** 2 L 560 ml = _____ ml
1/4 L = _____ ml	1300 ml = _____ L _____ ml

9. Answer the questions.

a. What months could you go sledding?

b. Find the three coldest months of the year.

c. What is the difference between April and July minimum temperatures?

Month	Minimum temp. (°C)
Jan	-37
Feb	-32
Mar	-15
Apr	-5
May	4
Jun	12

Month	Minimum temp. (°C)
Jul	18
Aug	16
Sep	-4
Oct	-20
Nov	-29
Dec	-34

Cumulative Revision, Grade 4, Chapters 1-6

1. Choose the number sentence that fits the situation. Solve for x. Think: what does x mean in the situation?

a. Dana had $18 and then earned more money. Now she has $33. $\$18 + x = \$33$ OR $\$18 + \$33 = x$	**b.** The meal at the restaurant cost $86 and Dad paid with $100. $x - \$86 = \$100$ OR $\$100 - \$86 = x$
c. Carmen dropped ten dozen eggs and all but 39 broke. $120 - 39 = x$ OR $x - 120 = 39$	**d.** The animal shelter found homes for 13 dogs in one day, and then had 43 dogs left. $13 + 43 = x$ OR $43 - x = 13$

2. Divide. Check each answer by multiplying.

a. $3\overline{)1\ 7\ 7\ 0}$ Check:

b. $3\overline{)7\ 9\ 0\ 2}$ Check:

3. Divide. Indicate the remainder.

a. $53 \div 10 = $ _____ R _____	**b.** $48 \div 9 = $ _____ R _____	**c.** $29 \div 3 = $ _____ R _____
$31 \div 11 = $ _____ R _____	$44 \div 12 = $ _____ R _____	$100 \div 9 = $ _____ R _____

4. Estimate by rounding one or both factors. Do not round both if you can calculate just by rounding one!

a. 7×78	**b.** 13×67	**c.** 311×8
$\approx$ ___ $\times$ ___ = ___	$\approx$ ___ $\times$ ___ = ___	$\approx$ ___ $\times$ ___ = ___

5. Add.

   a. $60\,000 + 70 =$ _____

   b. $123\,000 + 4000 + 4 =$ _____

   c. $3 + 90\,000 + 40 =$ _____

   d. $7 + 20 + 632\,000 =$ _____

6. Convert between measures of length.

a. 7 m = _____ cm	**b.** 2 m 6 cm = _____ cm	**c.** 4 km 100 m = _____ m
69 mm = ____ cm _____ mm	6 km = _____ m	169 cm = ___ m _____ cm

7. Convert between measures of weight.

a. 3008 g = ____ kg _____ g	**b.** 7 kg 940 g = _____ g	**c.** 45 310 g = ___ kg _____ g
4 kg 11 g = _____ g	4900 g = ___ kg _____ g	36 kg 140 g = _____ g

8. Solve.

a. While on vacation, the Smith family paid for different hotels this way: $345 for one night, $385 for one night, $370 for one night, and $320 for one night. What was the average cost per night?

b. Jessica walked around a square-shaped park. Each of the sides is 650 m long. How long a distance did Jessica walk? Give your answer in kilometres and metres.

c. Amy bought 3 pencils for $1.95 each, a sharpener for $2.10 and an eraser for $2.95. She paid with a twenty-dollar note. How much did she spend?

How much was her change?

Cumulative Revision, Grade 4, Chapters 1-7

1. Estimate by rounding the numbers to the nearest thousand or to the nearest ten thousand. Then calculate.

a. $22\,934 + 5312 + 424\,787$

Estimation:

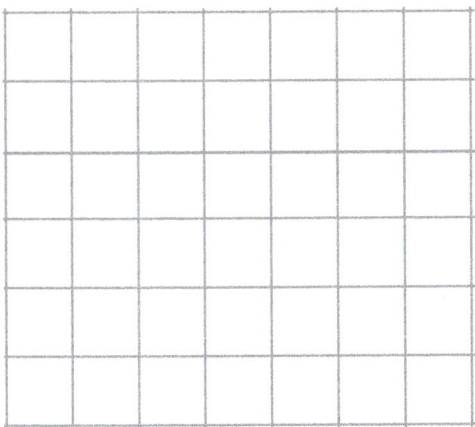

b. $519\,313 - 47\,616$

Estimation:

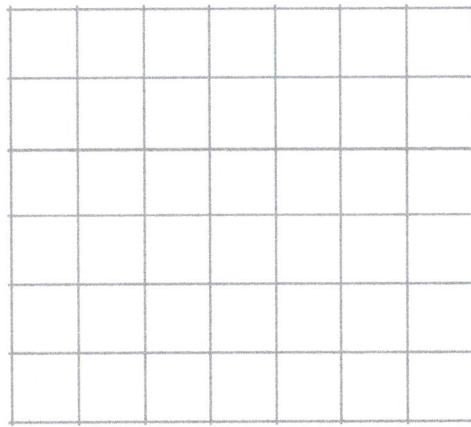

2. There are 45 students in each of the 22 buses, and 27 students in one additional bus. How many students are there in all the buses?

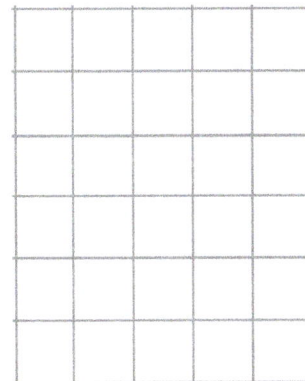

3. Fill in. Then sketch an example picture for each type of triangle.

Right angles are exactly _____°.

Right triangles have exactly _____ right angle.

Obtuse angles are more than _____° but less than _____°.

Obtuse triangles have exactly _____ obtuse angle.

Acute angles are less than _____°.

Acute triangles have _____ acute angles.

4. If the perimeter of a rectangle is 28 centimetres, then what could the side lengths be? Write possible side lengths in the table. Then calculate the areas. You can draw the rectangles in the grid, if you would like.

One side	Other side	Perimeter	Area
		28 cm	
		28 cm	
		28 cm	
		28 cm	

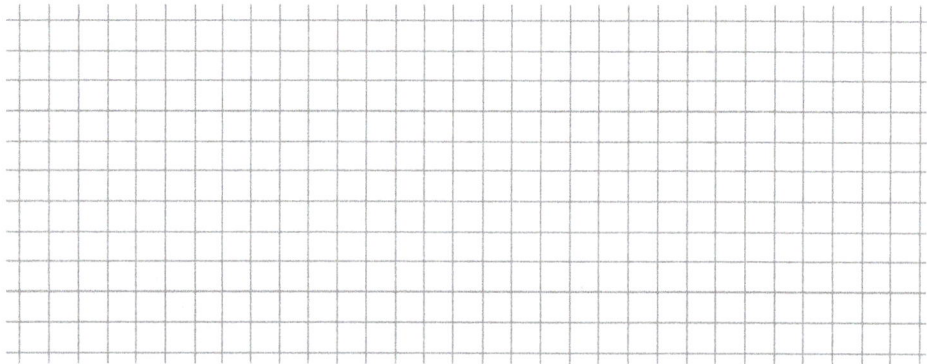

5. Change the times to the 24-hour clock times.

a. 1:40 p.m.	**b.** 9:20 p.m.	**c.** 2:15 p.m.	**d.** 10:04 a.m.
_____ : _____	_____ : _____	_____ : _____	_____ : _____

6. Mark an "x" if the numbers are divisible by 2, 5 or 10.

number	divisible			number	divisible			number	divisible		
	by 2	by 5	by 10		by 2	by 5	by 10		by 2	by 5	by 10
478				1492				904			
540				3093				905			
255				94				906			

7. Is 549 divisible by 7?
   Explain why or why not.

8. Find all the factors of the given numbers. Think of writing the number as a multiplication in many different ways. Do not forget the number itself times 1!

a. 24	b. 66
factors:	factors:
c. 96	d. 75
factors:	factors:

9. Draw the liquid in the thermometer. Match the temperatures to the situations.

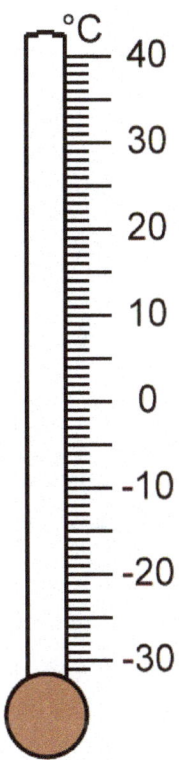 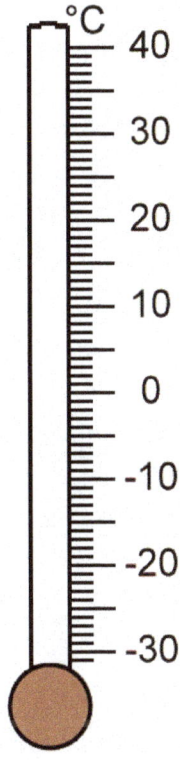 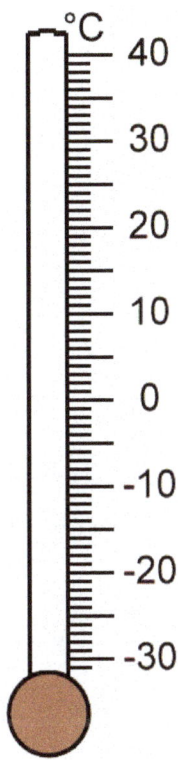 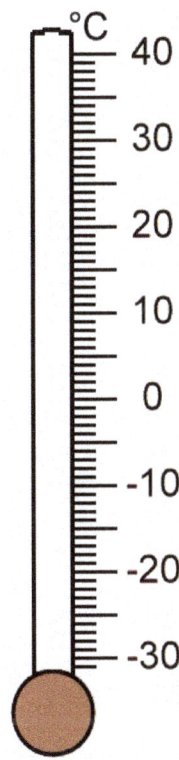 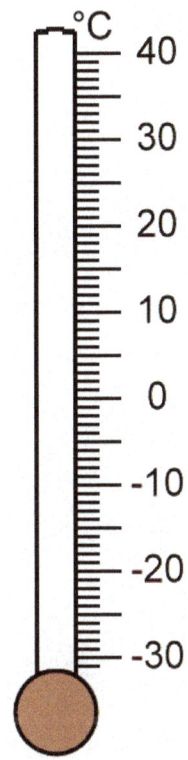

a. 10°C   **b.** 30°C   **c.** 20°C   **d.** 39°C   **e.** 0°C

a hot day   a chilly fall day   a winter day   fever   inside a house

Cumulative Revision, Grade 4, Chapters 1-8

1. Multiply in parts.

a. 4×36	**b.** 5×65	**c.** 8×426
_____ + _____	_____ + _____	_____ + _____ + _____
= _____	= _____	= _____

2. If you add 1 thousand, 1 hundred, 1 ten and 1 to this number,
   it becomes 100 000. What is the number?

3. Estimate the products by rounding one or both factors.

a. $8 \times 69 \approx$	**b.** $11 \times 55 \approx$	**c.** $25 \times 17 \approx$

4. Write the division problem. Solve for x.

a. The divisor is 8, the dividend is x and the quotient is 7.	**b.** The dividend is 24, the divisor is x and the quotient is 8.
_____ ÷ _____ = _____	_____ ÷ _____ = _____
$x =$ _____	$x =$ _____

5. Are the lines drawn symmetry lines for the figures?

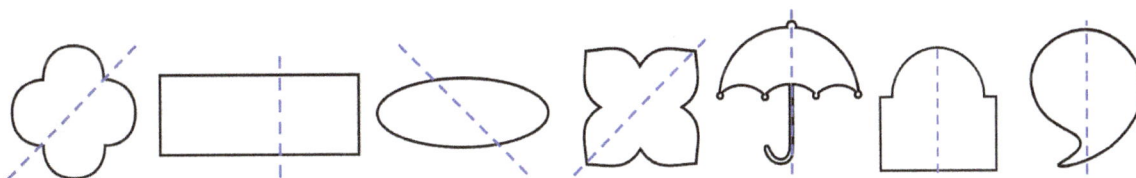

6. Convert.

a. 5 m = _____ cm	**b.** 3 m 4 mm = _____ mm	**c.** 4 m = _____ cm
12 m = _____ mm	6 m 6 cm = _____ cm	9 m = _____ mm

7. Underline the heaviest amount.

a. 5 kg 500 g 5050 g	**b.** 340 g 3 kg 400 g	**c.** 9 kg 9900 g 900 g

8. Solve the problems.

   a. Samuel weighed some kittens when they were two weeks old.
   They weighed 93 g, 93 g, 155 g, 62 g and 124 g.
   What was their total weight?

   b. Annie packed 175 kg of strawberries into 4-kg boxes.
   How many boxes did she need?

   c. Andrea used 1/4 of her $392 savings to buy balloons and other
   party supplies. How much money does she have left now?

   d. Thirty-six children are going to march in rows in a parade.
   How many children should be in each row so that the
   rows will be even?

   e. Two-thirds of the horses on a farm are full-grown and
   the rest are foals. There are 68 full-grown horses.

   How many foals are there?

   How many horses in total are there?

   foals

   ?

9. Draw angles of the following measures. Use a protractor.

a. 63°	**b.** 108°

10. Find rays, lines and line segments that are either parallel or perpendicular to each other. Use these notations: ‖ for parallel and ⊥ for perpendicular.

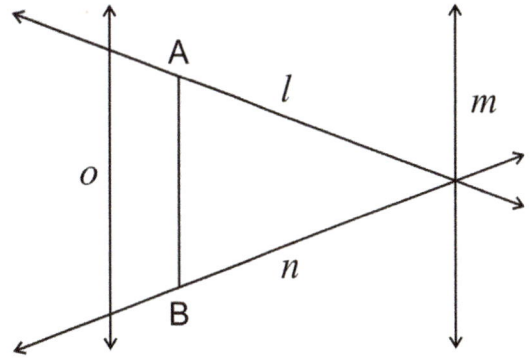

11. An apple was cut into 8 pieces. You ate 2/8 of it. How many fourths of the apple are left?

12. Add and subtract. Give your answer as a mixed number or a whole number.

a. $1\dfrac{2}{7} + 1\dfrac{5}{7} =$	**b.** $\dfrac{4}{12} + \dfrac{9}{12} =$
c. $\dfrac{1}{10} + \dfrac{3}{100} =$	**d.** $5 - 3\dfrac{3}{4} =$
e. $\dfrac{46}{100} - \dfrac{2}{10} =$	**f.** $2\dfrac{4}{10} + 3\dfrac{7}{10} =$

13. Multiply. Give your answer as a mixed number or a whole number.

a. $6 \times \dfrac{5}{10}$	**b.** $2 \times \dfrac{3}{5}$	**c.** $5 \times \dfrac{3}{4}$	**d.** $\dfrac{7}{10} \times 10$

14. Write the equivalent fractions. Shade parts in the pictures.

a. $\dfrac{4}{5} =$	**b.** $\dfrac{1}{3} =$	**c.** $\dfrac{2}{3} =$

15. Compare. Write < or > in between the fractions.

a. $\dfrac{3}{8} \ \square \ \dfrac{2}{4}$	**b.** $\dfrac{6}{8} \ \square \ \dfrac{8}{5}$	**c.** $\dfrac{2}{5} \ \square \ \dfrac{4}{7}$	**d.** $\dfrac{5}{10} \ \square \ \dfrac{5}{12}$

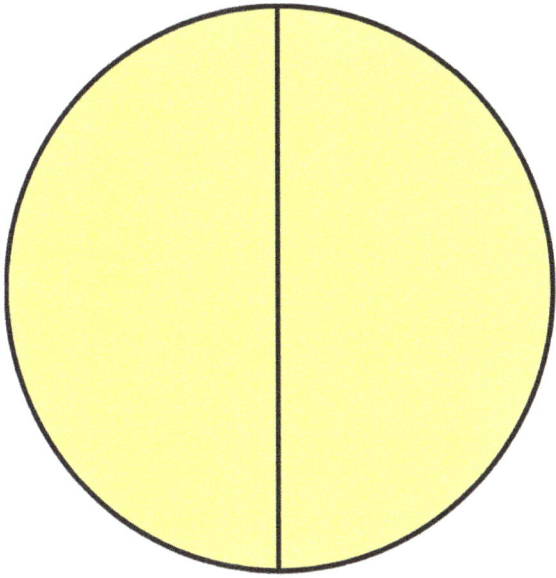
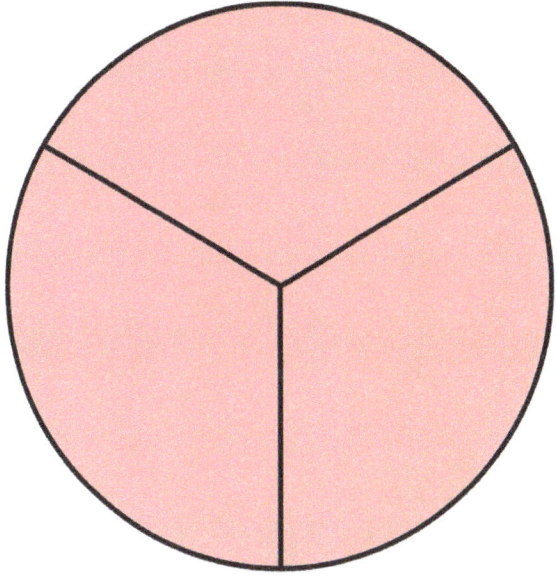
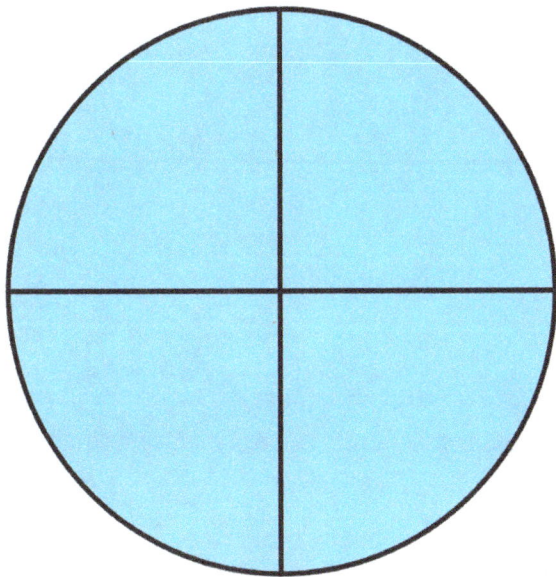
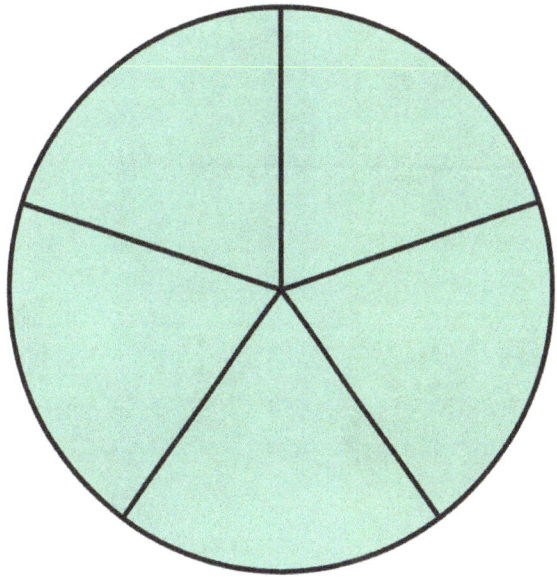

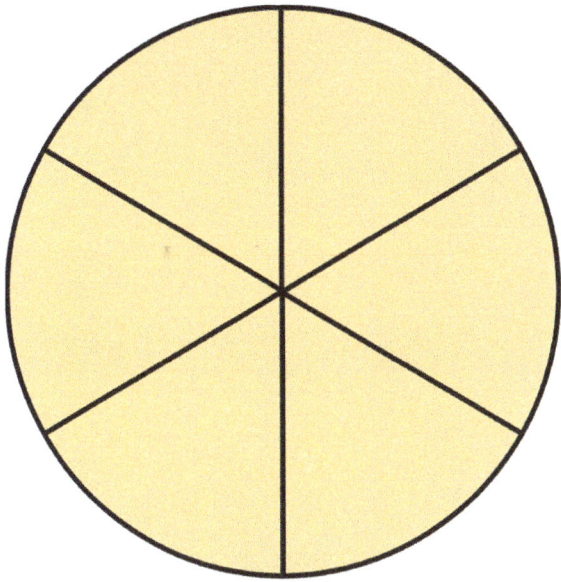
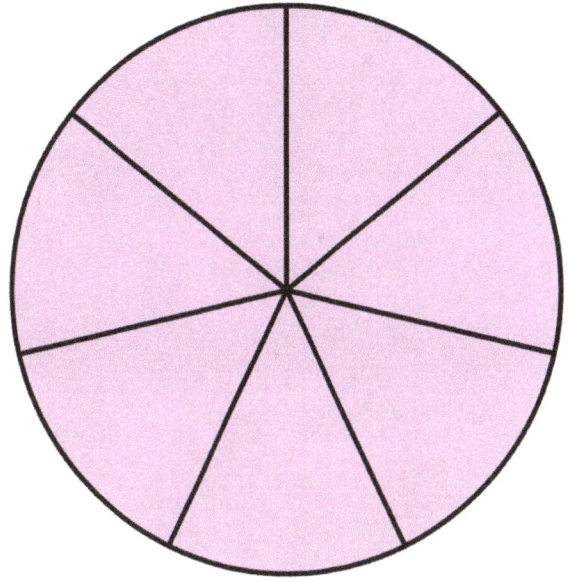
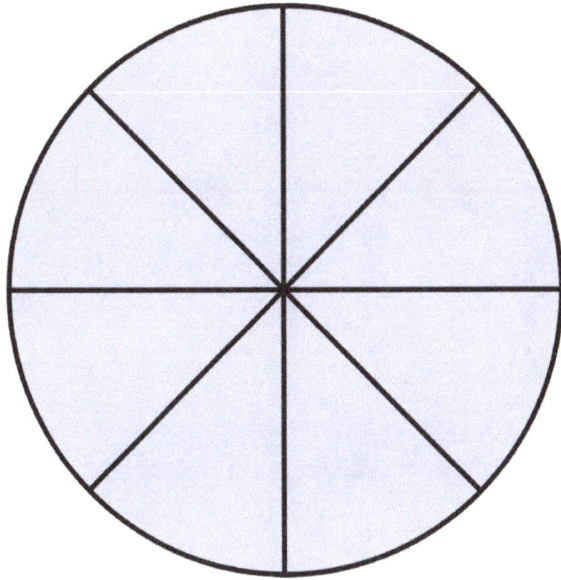
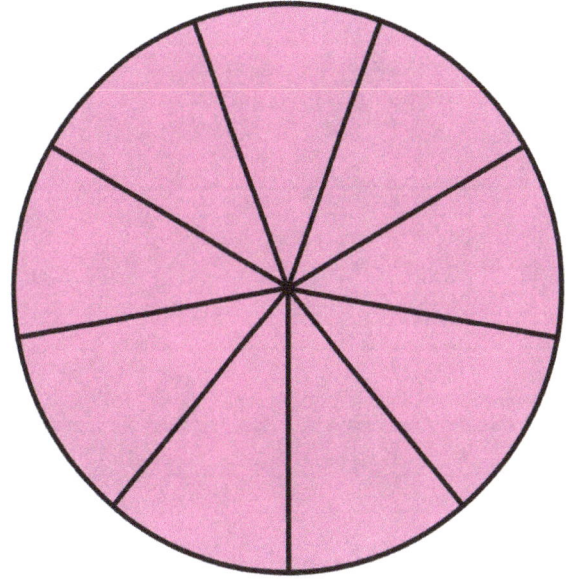

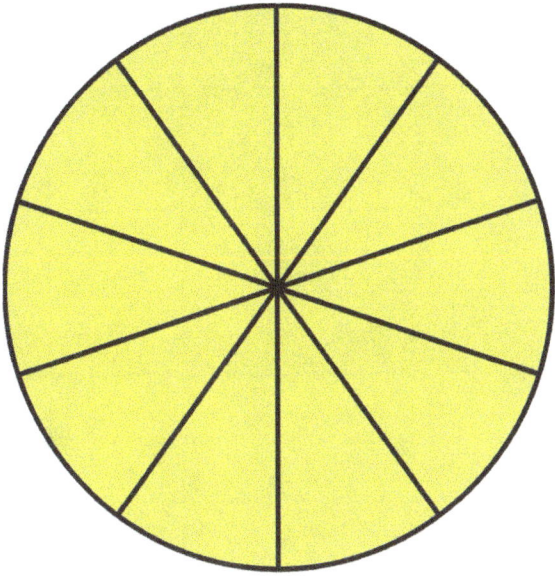
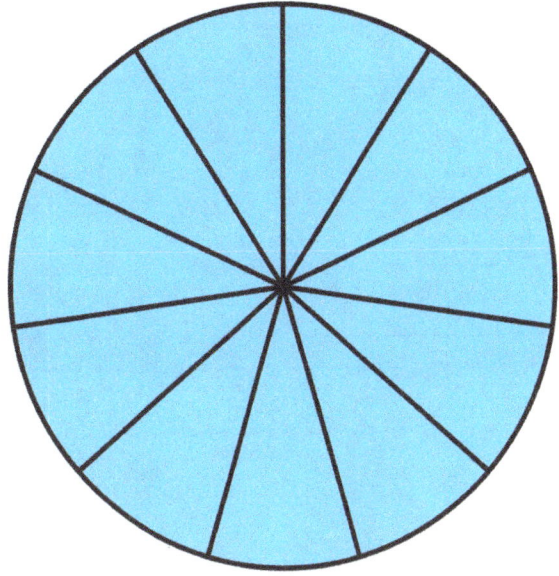
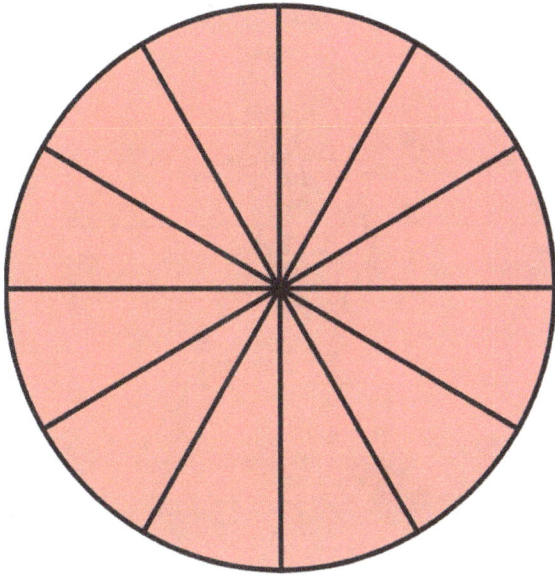
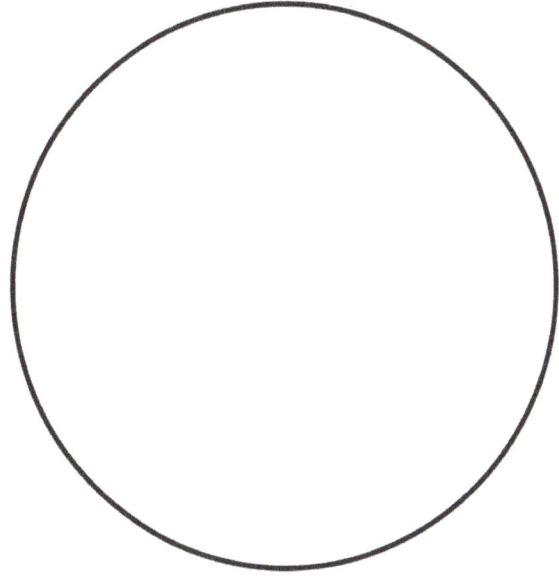

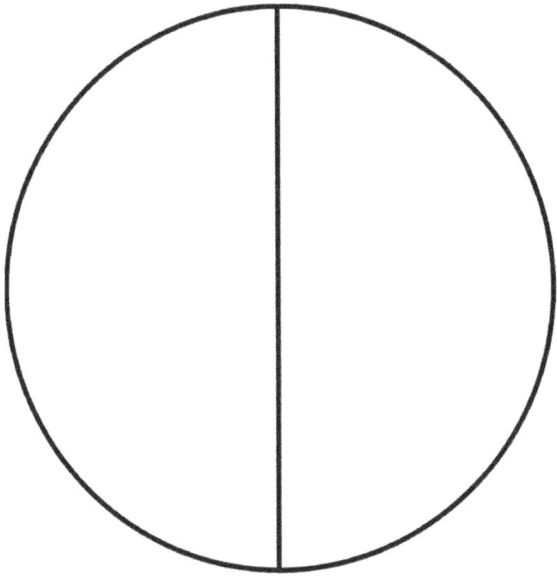

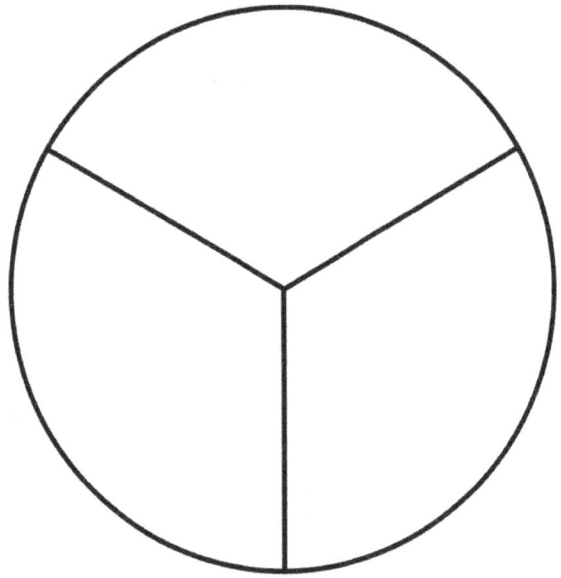

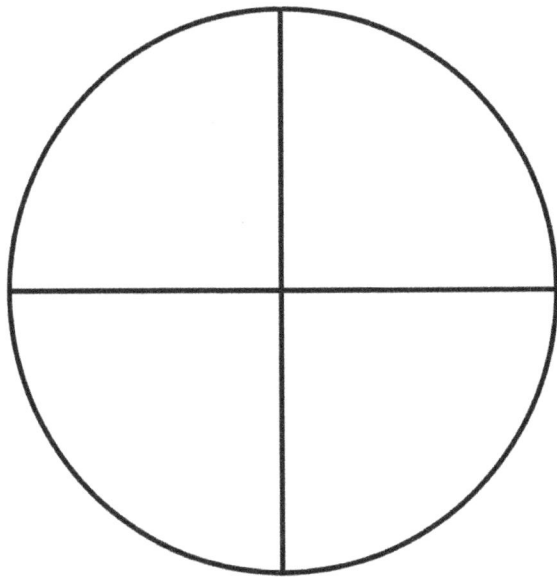

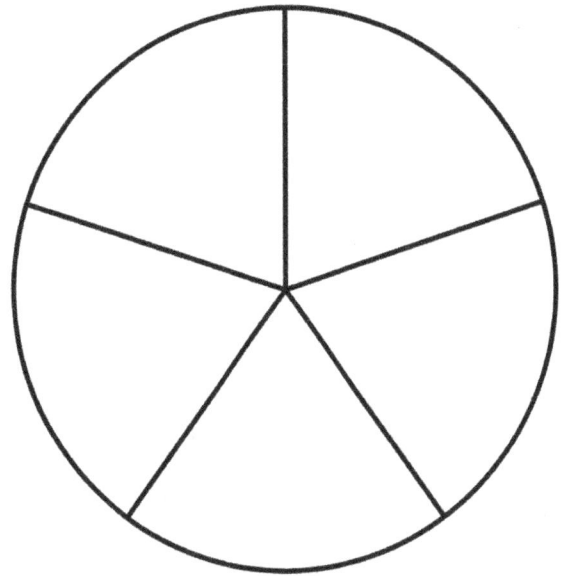

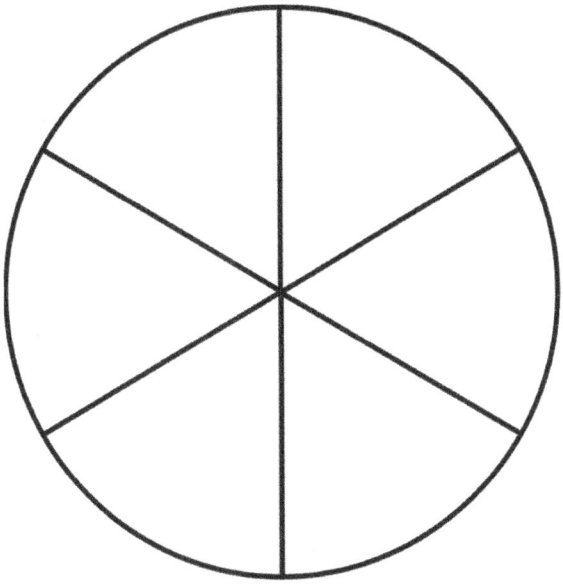
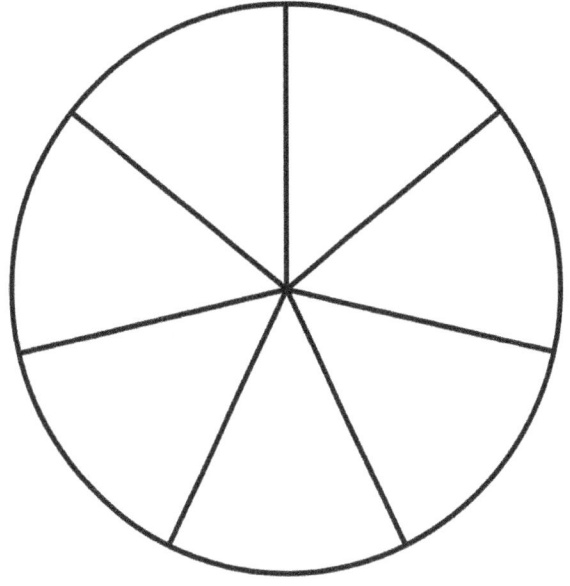
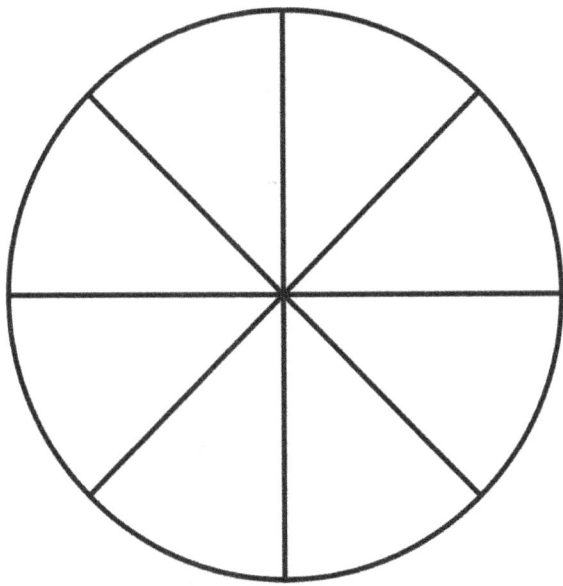
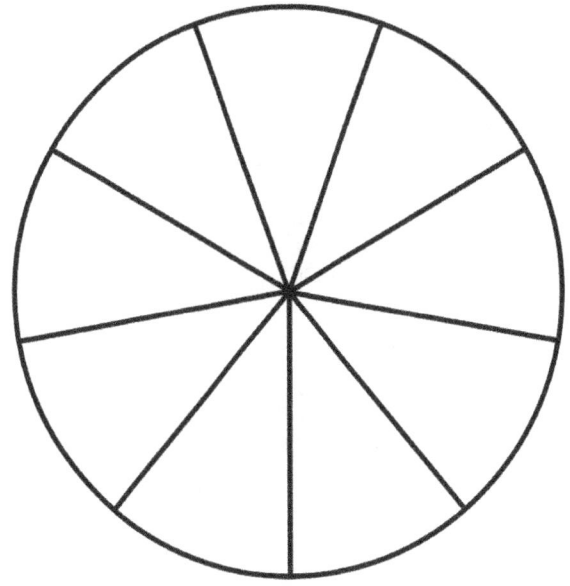

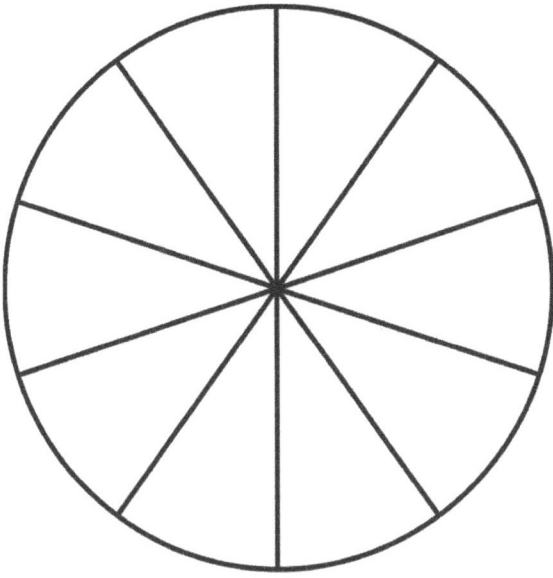
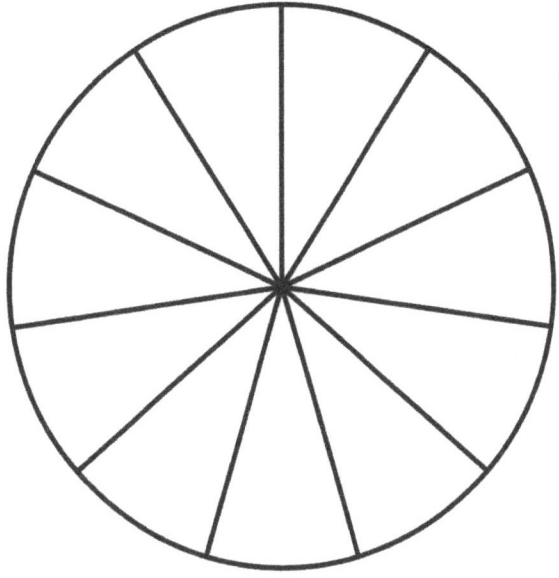
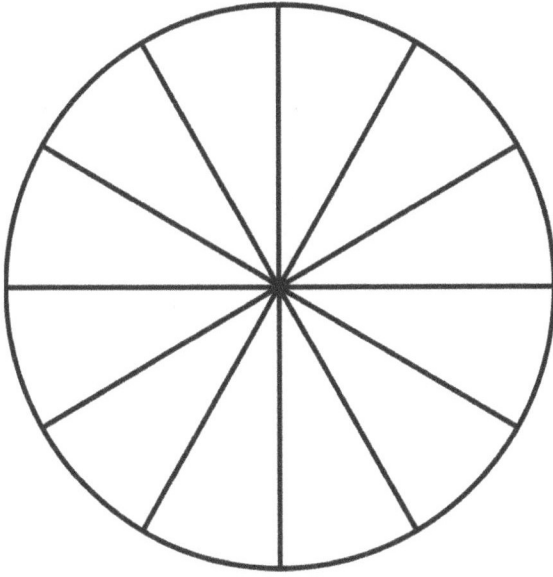
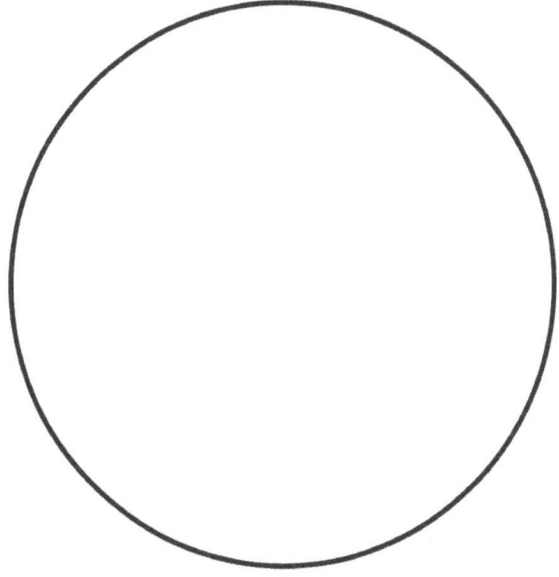